HOAs and ENVIRONMENT AND SAFETY and GOOD BOARDS AND BAD BOARDS

A Mini Study

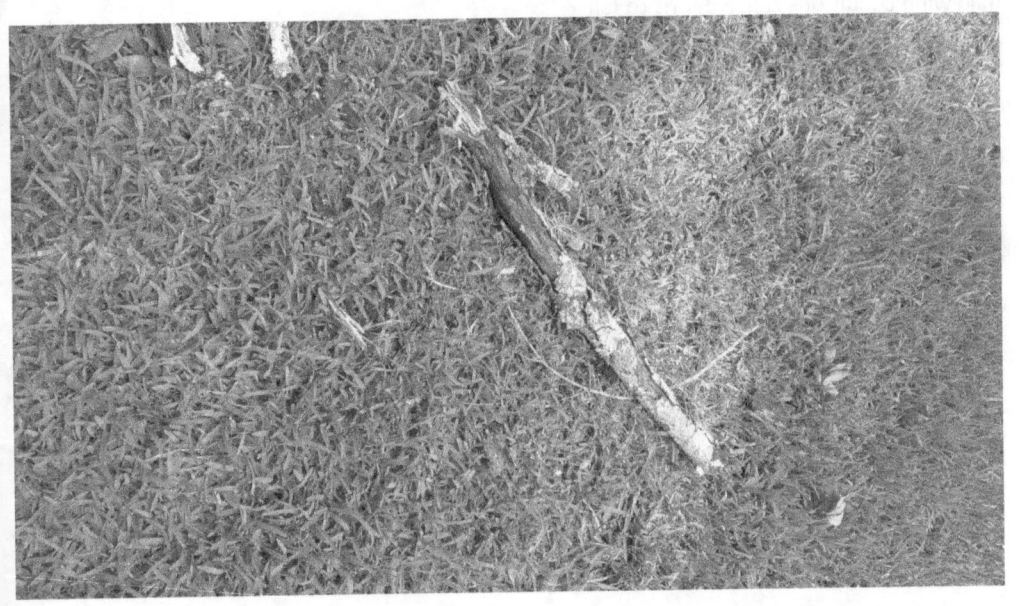

Arthur R. Maret Ph.D.

October 2019

I. INTRODUCTION

Scenario #1: An owner is mid-way through her daily walk around her subdivision. She has no walking impediments. Suddenly, she trips and falls. She gets a knee laceration, is bruised and her hip hurts. She manages to shuffle home. After about a week of recuperation, she is back walking – but never again will she walk on the sidewalk. Despite the danger, she walks in the road.

Scenario #2: An owner is working in the front yard when a large branch from the neighbor's aged oak tree falls and narrowly misses his head. Most of the trees on his street are in the same (untrimmed) condition. Large pieces of tree debris can be frequently found in the road - Even mild wind conditions cause debris to fall.

Scenario #3: An owner took a peek outside one afternoon and was surprised to discover that an enormous limb had fallen from the tree on the opposite side of the street. It looked as if fully half of the tree had collapsed. The limb was so large that ½ of the road was blocked off! Imagine the result if a car or person had been under the tree at the time of the incident?

Scenario # 4: A teenage resident is playing ball near an HOA's landscaping mower when she breaths a 'choking' cloud of black smoke emission from the mower. Several years later, although never a smoker, she dies of lung cancer.

Factual scenarios #1 - #3 have the same root cause: a lack of attention to safety issues.

Scenario # 4 is fictional example of the well-proven connection between vehicle emissions and the incidence of cancer. It is quite possible for a single exposure of a carcinogenic agent to cause cancer.

Environment and Safety (E&S) are crucial elements in the operation manuals of mid- to large-size companies thanks to OSHA and the EPA. In small and ultra-small ('Mom & Pop') companies, E&S is either non-existent out of ignorance ... or disregarded for financial reasons.

Who protects John and Jane Doe as they take a walk though their HOA-governed neighborhood?

Right now, there are few forward looking associations that actually factor E&S issues into their operational models. Frankly, most HOA boards could care less about E&S. "It's not in our charter", they say. Wrong!!

Author's Note: *He has both theoretical knowledge and practical know-how in the E&S area.*

II. E&S and HOA RESPONSIBILITES

Reproduced below is an extract from a particular HOA's Articles of Incorporation:

'The Association is formed for, among other things, ..., and to promote the **health, safety and welfare** of the Owners.'

To the Boards of HOAs everywhere: You need to step up and take responsibility for - AND control of - the E&S issues under your jurisdiction!

III. E&S IN HOA COMMUNITIES

In general, E&S can be sub-divided into several dozen categories, but for our purposes it is convenient to consider only six areas:

1. Safety Issues:

- Road Hazards
- Sidewalk/Driveway Hazards
- Tree Hazards
- Contractor Hazards

2. Environmental Issues:

- Air Quality Hazards
- Water/Chemistry Hazards

IV. SAFETY ISSUES

1. Road Hazards

In a HOA community, the roads may or may not be the responsibility of the association. If they are under association control then any injury or damage resulting from a pothole, for example, would render the association legally liable.

What is the situation with a city or county-owned right of way? Well ... if the Property Manager (PM) does a walk through the neighborhood – as SHOULD BE PERFORMED ONCE A MONTH – and the next day a car collides with a trash container that has been left in the curb area for several days ... guess who is liable? The owner, PM AND the association! (Remember, in most cases state and local government entities cannot be sued).

It should be obvious to the reader that readily seen dangerous road circumstances that are observed by a HOA board member should be noted, reported to the appropriate authority and then FOLLOWED UP. In a 'Good' HOA, there is a special committee that deals with safety issues. Some associations use the (already overworked) ARC for that purpose.

An optimum approach not just for road hazards, but for all safety issues, is to take proactive steps that help <u>prevent</u> an accident from happening. Conscientious, responsible parents take proactive steps every day – such as checking that the safety gate is securely latched at the top of the stairs so that their toddle daughter will not fall down the stairs. All owners should so the same *outside* their residences.

Board members: How many times has your BOD, or <u>you personally,</u> been proactive in a HOA safety area? This writer's experience over several decades has been that most board members have little interest in these matters.

Let's look at a (no-brainer) list of potential road hazards:

- Potholes
- Height difference of adjacent parts of the road – hazard for bike riders
- Low lying or sunken areas that collect standing water after a storm
- Flash freezing of ice in low lying road areas
- Missing or mis-placed road signs
- Traffic signals non-functional or badly timed
- Lane and other painted markings faded or worn to the point of illegibility
- Need for 'speed bumps' to help control speeding
- Trees/bushes/landscaping impeding visibility at a junction
- Utility structure impeding visibility at a junction
- HOA contractors using landscaping equipment and golf carts in a careless manner

A couple of the above areas deserve a more detailed discussion:

1) Freezing of water in a sunken road or sidewalk area is a major winter hazard in in southern states, and the cause of even fatal falls and accidents.

2) It is amazing that some HOAs who do not waste a moment citing a homeowner for the wrong color paint on the front door stand by month after month and do nothing about landscape that impedes road visibility! This is an often-ignored hazard. This issue underscores the need for FREQUENT and DETAILED INSPECTIONS WITH PHOTGRAPHIC EVIDENCE by the HOA.

3) It is not uncommon to find a large electrical transformer box located at an 'inconvenient', location-blocking visibility at a road junction. A little pressure from a motivated HOA might just get the structure relocated a few yards distant, thus improving visibility/safety. The photos below show a typical stupid situation that should have been fixed years ago: "I need to inch forward a little more to see the traffic to the right … no … still further … Oh … But I'm now beyond the official stop line and I'm completely over the pedestrian crossing!! … Just a bit further! …"

4) HOA contractors are not known for being the most safety-conscious companies/employees – and <u>HOAs and PM companies are not known for being among the most stringent for compliance with OSHA/EPA/Common sense rules, either!</u> This area is truly a quagmire of danger for the residents of a HOA community. Let's look at some REAL safety issues:

a) Use of motorized golf carts and 'dune-buggy-type' vehicles on public roads and sidewalks.

- Has the local sheriff given permission for these vehicles to be used?
- Do the drivers/crew get ANNUAL training and driving check-ups?
- Are the drivers given a regular sobriety and drug test?
- What is the maintenance schedule for the vehicles?
- Does the sheriff do an annual compliance review of ALL drivers and ALL equipment?
- Does the HOA have an equipment safety inspection done annually by a mechanic?
- Are the brakes tested weekly?
- Does the HOA have an operations manual for both the vehicles and the drivers?
- Does the contractor post a warning sign or orange cones in the areas of work?
- Are there special procedures to be followed when children are present?
- If the vehicles are powered by a prime move, i.e., a gas engine, are the emission levels within state/federal limits, and are the emission levels checked on a regular interval

Of course, most of the above issues also apply to regular landscape equipment such as motorized mowers

b) General Issues

- Does the HOA have MSDS sheets on file?
- Does the HOA have an operating manual for safety?
- Are contractor employees certified?
- There are too many etc., etc., to keep listing these no-brainer items

2. Sidewalk/Driveway Hazards

The unfortunate 'trip-and-fall lady' in the introduction is one of thousands of victims nationwide due to carelessness/maintenance by homeowners and LACK OF ENFORCEMENT by HOAs.

Uneven pavement is a major issue for housing developments in 'old age'. If there are lots of trees nearby ... then say goodbye to having a level driveway or sidewalk!

ARC folks – who are charged with EVALUATING ALL VISUAL ASPECTS of a neighborhood – are FAILING in their responsibility to note and cite safety problems that can readily be observed.

This section of sidewalk has dropped 1 ½ inches = **MAJOR TRIP/FALL HAZARD**

Another safety problem (more common than you would think) is a sidewalk that disappears into oblivion , thus forcing children to WALK IN THE ROAD NEAR A BUSY INTERSECTION TO CATCH THEIR SCHOOL BUS!!!

3. Tree Hazards

The photo on the cover page of this report shows a rotted tree limb FOUR FEET LONG and weighing over 7 pounds. Imagine the result if this fell on someone's head!

There are a few persons who hate trees to the nth degree. But, most love trees – even if leaf collection is a chore – and a few extremist complainants call their local HOA regarding a tree that is being legally trimmed by a neighbor!

Trees shed branches from time to time and trees, like humans, get old. Any tree greater than 10 years old, especially a hard wood variety such as oak, is a source of great danger for residents. Why? Because the tree will already have substantial insect damage to the critical heartwood (central) area. A main trunk or branch of a tree is not held in place against gravity by the bark … it's the inner fibers that serve to keep debris from falling. Another factor that causes a weakened limb is wood rot – more properly called fungal attack. *Armillaria rot* is a particularly nasty organism and can attack an entire forest and collections of trees in HOA sub-divisions. So … even a middle-aged tree can have hidden health problems that render it susceptible to collapse. What about an 'old' tree? It's much worse – limbs fall off due to the effect of gravity on the weakened limb. Limbs fall without warning as documented in the two cases mentioned in the Introduction.

A HOA community that is built on poorly drained, clay-type soil has an additional problem: Exposed tree roots – which themselves help to push the surrounding soil upwards … leading to cracked and irregular concrete driveways and sidewalks. The exposed roots make the tree VERY SUSCEPTIBLE to falling.

A 30-year old HOA community that has thousands of old trees certainly is in a tough situation:

- Is each tree pruned on a regular basis (You know the answer is "no.")
- Has each tree been recorded on a spreadsheet? (You know the answer is "no.")
- Is each major limb that falls documented? (You know the answer is "no.")
- Has each tree been evaluated for safety concerns? (you know the answer again).
- Does the HOA have a state certified arborist as a consultant?

If I were the parent of a child who is required to walk to school underneath trees with known falling limbs … no matter how small … I think I would be very upset with the HOA. Board members – please take the time to look at just one old tree in your neighborhood. If you see fallen small branches every few days … then the tree is dangerous to walk under. It's that simple. You see – the Board of a HOA has a solemn responsibility to protect the residents just, as an owner is obliged make sure his front door is painted in the approved color.

Those tree lovers who want to preserve an old tree in an HOA despite the obvious dangers will – in 10 years – be paying exorbitantly high HOA fees, and 80% of the budget will be devoted to dealing with an 'out-of-control' situation.

HOAs that were chartered more recently have learned an important lesson from their predecessors: 1) They don't plant many Magnolia trees anymore; 2) Trees are pruned on an annual basis; 3) Trees older than 10 years are evaluated for replacement; 4) Trees are NOT planted at 50 ft. intervals, or closer; 5) Trees are treated with a fungicide on a regular basis.

One might believe that this policy results in higher dues to the owners. It's actually *less* expensive!

This is an example of 'cost-benefits analysis' at work ... a skill that someone on a HOA Board should have ... and the PM company certainly SHOULD POSSESS.

4. Miscellaneous Hazards

Look at the photo below. If you cannot discern a safety problem that needs to be addressed by the HOA ... then I guess I have wasted my time with this white paper ...

V. ENVIRONMENTAL ISSUES

Hopefully, every HOA board member is aware of the general issues that surround air and water purity/pollution. What about noise pollution?

1. Air Quality Hazards

Air quality is a complex situation. Why? Because air pollution is not merely one ('idiot') number for a city such as Tampa ... Air pollution is a 'micro-environment' parameter that can change in a space of only a few yards! Three factors that we will discuss are:

- Auto emissions
- Emissions from HOA/Contractor landscape equipment
- Dangerous pathogens

a) We cannot do much about auto emissions in the neighborhood except to hope that the 20-year old car next door doesn't develop a leaky engine gasket.

b) The second listed air pollution route – HOAs and contractors are, in fact, **VERY SIGNIFICANT** sources of local air pollution. Let's look at the three main factors:

- Engine-driven lawn equipment have very high emission levels (no catalytic converter).
- Poor/zero maintenance = even worse emissions.
- Often poorly trained and supervised operators.

If ever a person should be awarded a negative Nobel Prize ... it should go to he/she who invented the gas-powered blower. That device alone accounts for more than 50% of emissions generated a typical landscaping company – and the noise 'pollution' it creates – well it's too ridiculous to bother with a tirade. All too often landscapers merely blow grass cuttings, leaves and small debris onto the common area - or onto another community's land! HOAs are generally too cheap to pay for proper collection and disposal. We wish that HOA Boards were aware of the 'entropy effect' that is caused by shoddy disposal practices.

Trees – again -To those who love an avenue of densely packed trees that overhang the road. Think of this: You are walking in a micro- environment that is up to 10x more polluted than a road with fewer trees, if the landscaper has recently paid a visit with his leaf blower ... and the air quality is poor in any event due to the passage of autos!! HOAs: Have the air/emissions tested!

c) Pathogens

We live in an ocean of air. And ... within the lower atmosphere are zillions of live and dead pathogens of various types. Unlike the Martians in 'War of the Worlds' humans have genetically adapted to survive quite nicely alongside the majority of airborne 'bugs'. However, some pathogens that can be transmitted in the air are so deadly that just a few spores can cause death.

14

This author spent many years on a Government program on chemical and biological warfare agents: *Bacillus anthracis* and *Ricinus communis* are deadly to any animal species! Anthrax spores can lie dormant in soil FOR YEARS and are released in the atmosphere by any process that disturbs the soil ... and that includes just pulling deep rooted weeds!

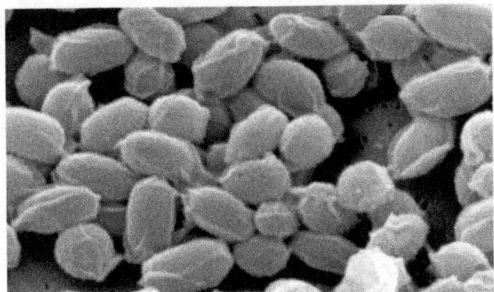

Bacillus anthracis

What about the mold that appears on sidewalks, roofs and shaded parts of a home? Is this stuff bad? Well, some persons are very allergic to molds and in fact can suffer severe illness. There are hundreds of different fungal species that create what we see as a mold. The common black fungus/mold, *Aspergillus Niger*, is tolerated by most persons when inhaled – but again susceptible persons can get sick from the exposure. Lurking amongst black mold is often the much more lethal *Aspergillus Fumigatus* – a green looking substance ... and this mold can appear on its own both outside and inside homes. Aspergillus Fumigatus is VERY dangerous. This pathogen can settle in the lungs of persons with a poor immune system (children, elderly, cancer patients, etc) and cause **aspergillosis**. This is a very debilitating fibrous lung disease ... and can appear several years after the initial exposure!

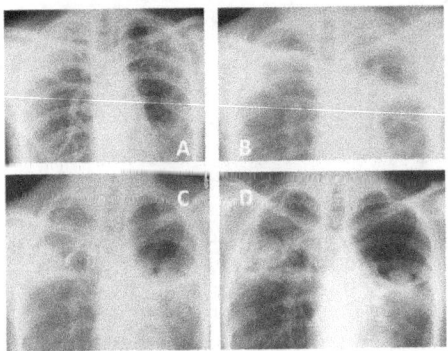

Pulmonary aspergillosis

TREES ... and more trees. It's obvious that the presence of trees that overhang sidewalks and gardens help molds to flourish!

What can be done by an HOA to help minimize the exposure of their residents to dangerous pathogens? Many associations do nothing! HOAs with a high tree count SHOULD BE VERY PROACTIVE IN MOLD CONTROL – even more so in high humidity states in the southern US.

Those HOAs with a more 'picky' perspective on appearance are unwittingly helping the health of their residents as follows:

- Some associations require roofs to be cleaned when visible mold/mildew occurs
- Some require sidewalks and driveways to be cleaned of mold
- A few require mailboxes to be cleaned of mold

The writer has lived in a community where all three of the above requirements were in place.

Board member: What is your association doing about minimizing the health effects of mold?

Heavy sidewalk mold in the writer's community

Extensive driveway mold in the writer's community

2.Water/Chemistry Hazards

Since most HOA communities are served by piped water, it is often 'assumed' that the water quality is OK. That was obviously not a good bet in Flint Michigan!! In fact, many hundreds of communities in the US have water supplies that do not meet all EPA specifications. Someone on the HOA Board should be required to review the annual EPA water report. The water quality should be an agenda item at a BOD meeting.

- A HOA community with a clubhouse obviously uses the public supply. What steps does the HOA or PM take to ensure the safety of the water given that most faucets are not in hygienically clean circumstances (children/visitors/maintenance persons, etc)
- If re-cycled water is used for irrigation ... what safety notices/precautions are taken to protect children and animals?
- HOAs are among the biggest users of fertilizers and pesticides. What steps has your HOA taken to minimize the impact on groundwater contamination and harm to persons and pets? Has the plan been approved by an expert?
- Is someone on the Board charged with the responsibility of reviewing MSDS sheets, and monitoring the activities of contractors?
- Does your HOA have a rule in place that forbids the dumping of chemicals (such as motor oil, acids, paint washings, etc.) in the soil?
- Does your HOA's Website have a section that deals with hazardous waste policies?
- Who is monitoring retention ponds? What are the numbers for dissolved oxygen and nitrates, for example?
- Is there a plan to deal with the health effects resulting from hurricane flooding or melt from a massive snowstorm?
- Is the HOA's common property graded to minimize standing water?
- What mosquito control policies are in effect?
- What special precautions have been taken – including notices on the Web site - regarding the dangers of the Zika virus and Lyme disease?
- As a Board member ... do you know the basics of water chemistry? (Everyone should!)

VI. PERSPECTIVE

We are all exposed to environmental and safety hazards in our community and in the workplace.

The US government has taken the lead and formed several departments to deal with these issues, and all citizens have a responsibility to both comply with the law and use common sense.

What, then, is the role of the HOA in helping optimize the environmental 'quality' and increase the level of safety in a neighborhood? Or – perhaps more important – what is the HOA's responsibility to NOT make pollution/safety WORSE? We again reject the responses of those Board members and HOA attorneys who say that safety and the 'environment' are outside the purview of the HOA. What a stupid, irresponsible position!

Based on my many years of HOA experience, I have yet to see ANY Association that has completely fulfilled its solemn obligation to promote the health, safety and welfare of the Owners. In particular, most HOAs:

- Do not have a safety committee
- Do not perform safety inspections
- Do not vet potential and current contractors for compliance with OSHA, etc
- Do not discuss, proactively, potential safety enhancements at a Board meeting
- Do not rely on the services of a Safety Expert – for example, do you know what a person with the credentials MSPH, CIH, CSP behind their name does for a living?
- Wrongly assume that the Property Manager is fully up to date in E&S areas and relies solely on the PM to alert the Board of any issues

Yes .. as with many other problems with HOAs … The PM company must take a large share of the blame for the poor performance of Associations in the E&S area:

- Small PM's are struggling just to make ends meet – and focus mainly on new business - and in any event have few staffing resources to really research E&S issues
- Ultra-large PM's are often so caught up with the thrill of their success – and given that **few PM staff have ANY scientific training/interest** – E&S is given token whitewash attention … and the hapless OWNER is expected to happily rely on whatever edicts (even if clueless) are proposed by the PM to the Board of the Association

This mini study has highlighted potential very serious Environment and Safety deficiencies in the management of HOAs. Persons and animals have, and will continue to be, harmed by the lack of attention of HOA Boards to E&S. Yes … it's impossible to achieve a perfect score in the E&S area. But, HOAs should at least TRY to set a strategy to optimize owner health and safety.

(See next page)

20

VII. APPENDIX

A GOOD HOA BOARD:

- Has a member with background/interest in E&S issues
- Has a definitive strategic plan for E&S – and updates the plan on an annual basis
- Promotes health and safety in various communications with the owners
- Inspects for E&S problems on at least a monthly basis
- Has a data base system for tracking and reporting E&S issues
- Reports E&S issues at each Board meeting and documents fixes/actions
- Reports E&S issues on the Website and documents fixes/actions
- Has an emergency hot line for reporting ANY suspicious E&S problem (NOT merely an 'after hours' number)
- Has a committee dedicated to E&S issues
- Meets with law enforcement and county environmental staff at least annually
- Has a file of MSDS sheets for ALL substances used on the property - both in the Association's official electronic records - and on the Association's Web site.
- Has read and remembers the basics of OSHA and EPA regulations
- Places a standard E&S and OSHA/EPA section in every contract executed with vendors
- Is proactive in ensuring that the infrastructure of Board members' homes is free of mold, sidewalk imperfections, etc

FINALLY ...

A Good Board Focuses on Serving PEOPLE FIRST ...

and PROPERTY/APPEARANCE INFRACTIONS SECOND

GOOD HOA BOARDS AND BAD HOA BOARDS

A Mini-Assessment Report

Arthur R. Maret Ph.D.

October 2019

I. Introduction

FL is the Condo and HOA capital of the USA – 27,000 condo associations and 13,000 HOAs! Unfortunately, Chapter 720 of the FL Statutes is regarded as one of the weakest HOA laws when compared to other states. This situation, coupled with a seemingly perpetual limited staff at the department assigned to evaluate complaints (DBPR), has allowed unscrupulous developers, HOA boards and property managers to wreck homeowner lives in an almost carte blanch fashion. Look on the internet: every conceivable area of HOA 'life' can be found as a litigation topic or complaint thread on one of the many HOA 'chat rooms'.

Fortunately there are a few dozen law firms in FL that specialize in condo/HOA affairs. These firms get most of their income from associations. But, they often run a web site tailored to homeowners. Many of these sites provide generic answers to owner questions, and also give late-breaking news on changes in the law or regulations.

They SHOULD ... But very few prospective buyers of existing homes actually DO ... evaluate the 'quality' of the HOA, and - much more important - the level of experience and reputation of the property management (PM) company. They only discover the deficiencies after they have signed the mortgage papers. Did the seller disclose that the HOA was sub-par, or worse - useless? Of course not!

Assuming the prospective buyer was astute enough to actually probe the operational characteristics of the HOA, what questions should he/she ask?

Thanks to the internet, one can get many legal and layperson viewpoints on the criteria for a quality HOA. An HOA is truly US, THE OWNERS! But it's the Board of Directors (BOD) who must take all the blame for any deficiencies.

In the next section, I have reproduced a list of characteristics for 'good' and 'bad' boards. This list was compiled by a California law firm. In a later section, I present some additional characteristics of 'good' boards, based on my own experiences with 13 different HOA/Condo boards.

II. Good and Bad Boards

A Good Board:

1. Communicates with owners and residents on a regular basis, explains its decisions, openly discusses problems and victories, has a policy of transparency and truthfulness. Postings on bulletin boards accessible to all residents are key in this respect. Information meetings may take place occasionally.

2. Addresses residents' legitimate complaints/concerns/requests and respects useful suggestions.

3. Follows and enforces rules consistently and for everyone: Board members have to follow rules themselves if they expect others to follow them and should not show favoritism.

4. Exercises due diligence regarding contracts for repairs, landscaping, and staffing. In other words, a good board seeks tenders. When maintenance problems arise, a good board not only seeks advice from non-interested parties (to avoid conflicts of interest), but also asks if there is a better and less expensive solution than the one suggested by contractors.

5. Is constituted of members who have no axe to grind or a vested interest or a personal agenda.

6. Always respects an association's finances, assets, and owners' monies.

7. Makes certain that the common elements are well maintained and that the staff is competent and hard working.

A Bad Board:

1. Rarely communicates with owners on substantive issues and prefers to inform them as little as possible. This seems to be a key ingredient in a lowered quality of life, and is reflected in the many other problems that seem to accompany this issue.

2. Responds dismissively or angrily when owners justifiably complain to them about problems (such as broken rules) or unnecessary expenditures. Or, yet, simply ignores owners' concerns.

3. Threatens owners with legal action when they complain justifiably or make suggestions; or yet when owners complain about management, staff, and contractors.

4. Mistreats, harasses, threatens, or refuses services to owners who have justifiably complained or made useful suggestions.

5. Rubberstamps decisions made by the manager, administrator, superintendent or contractors without independently studying the issue. Does not get quotes for projects or services.

6. Spends monies for upgrades just to suit themselves, contractors, or managers.

7. Refuses owners' requests to view corporation records and documents.

8. Does not supervise manager and staff sufficiently. As a result, the work and services may be of lower quality or very little work may be accomplished. Or, yet, the staff is actually the power in the association.

9. Forms a clique, often with management, against owners, and fails to understand that a board represents OWNERS and not themselves nor the management/staff.

III. The Author's Addendum to the <u>Good</u> Board List

A Good Board:

A1. Is culturally diverse and has members with different workplace backgrounds.

A2. Operates akin to a 'good' private company: a) has a 5-year strategic plan that is updated annually; b) is proactive – stays abreast of current developments in FL HOA law; c) Gets COMPETITIVE bids EACH year for each major expense including the for the PM; d) uses a spreadsheet data base system to record, monitor and process key activities; e) performs a risk analysis at regular intervals of health and safety issues in the community.

A3. Manages the PM company: a) with a well-defined scope of work; b) by providing biannual performance review feedback; c) by meeting with the CEO of the PM at quarterly intervals; d) by carrying out an annual survey of owners on their views of the PM's performance.

A4. Supports the board President by a) performing assignments on a timely basis; b) offering un-biased views on issues, and does not allow the President use charisma or bullying to drive a decision in a certain direction.

A5. Applies an appropriate level and detail of systems analysis and decision-making techniques to ensure that critical issues get the best possible review before decisions are made.

A6. Has members that take the time to become acquainted with issues/problems in neighboring communities. They stay abreast of current FL HOA news by subscribing to on-line newsletters. (Such as HOAleader.com.)

A7. Has made the effort to digitize old records so that there is a complete computer-based chronological record of the association's activities.

A8. Does not canvas for a particular board candidate during the election cycle.

A9. Has members that are committed to making the association RELEVANT and USEFUL to the changing ages and lifestyles of a new generation of owners, while ensuring that the core values of the association are maintained/strengthened.

A10. Believes the association CAN and SHOULD be more than merely a compliance organization, as perceived by many owners nationwide.

IV. Perspective

After the reader has stopped chuckling – or maybe outright laughing – at some of the above criteria, let us now coalesce all of this author's thoughts into two main themes.

Many homes AND their HOA's in central FL are reaching middle-age status. All homes deteriorate with age – but wood frame houses in the FL climate are VERY susceptible to decline – as are the infrastructure elements such as trees. Up 'north' it is common to see wood frame houses that are 120 years old. How many 100+ year frame homes have you seen in FL?

It is a similar situation with homeowner associations. Developments that were built in the 1970's and 80's generally have unwieldy, poorly written governing HOA documents. A few (and it is only a few) forward looking associations have completely revamped their documents to 1) bring them into compliance with current FL law/regulations; 2) make them more relevant to the so-called younger generation that is now the predominant ownership and 3) laid out an operations pathway to create homeowner enthusiasm and interest that will sustain the HOA into the future.

All the above sounds like a good idea – at least in general terms. What is the biggest impediment to such a plan? You will be surprised when I say that it is the PM company! Sad to say, providing PM services to HOA's is now 'big' business. I have found that even a 'strong' BOD is heavily influenced by the PM and its law firm. Large HOA's out of necessity NEED a PM. But the PM role should be squarely in the support arena – and that's why a detailed definitive scope of work is so important.

In this author's view, there are no 'half-full' solutions to an optimum HOA – just black or white. Either the HOA BOD elects to continue to employ dated documents, has no strategic plan to meet changing circumstances, etc., or it enthusiastically embraces a 're-born' concept to optimize the HOA for the present AND the future.

I do not expect to get anywhere near unanimous agreement on my viewpoints, but I hope that this document will at least stimulate discussion on appropriate ways to optimize and sustain our HOA into the future.

GENERAL DISCUSSION:

A majority of the many millions of persons who live in a HOA community never read the governing documents, nor bother to vote or attend Board meetings. This has enabled their more active members not only to elect to the Board candidates with special interest views, but allow the Association to be managed in a dictatorial way that may or may not be beneficial to the long-term future of the Association.

General Discussion